LIZARDS
and
TURTLES
OF SOUTH-CENTRAL TEXAS

By

Thomas G. Vermersch

Living Jungle Science Programs
San Antonio, Texas

EAKIN PRESS ★ Austin, Texas

FIRST EDITION

Copyright © 1992
By Thomas G. Vermersch

Published in the United States of America
By Eakin Press
An Imprint of Sunbelt Media, Inc.
P.O. Drawer 90159   ★   Austin, TX 78709-0159

ISBN 0-89015-842-8

Dedicated to

Bessie Vermersch,

*my mother, who provided love, encouragement, and transportation
for many of my childhood field trips to observe reptiles,*

and

Joseph Laszlo,

*my friend, who passed away in 1987. He shared his remarkable
enthusiasm for herpetology and a unique and colorful view of life
which affected all who knew him.*

Contents

Preface

I have collected, observed, and studied most of the thirty-three lizard and turtle species and subspecies of South Central Texas at one time or another over the past twenty-six years. During this time regional data was collected on the distribution and the biology of these reptiles. I have incorporated some of this information into my regional field ecology and educational training programs for regional school districts.

This book was prepared after careful consideration of already existing state reference books on the subject with generally limited reference to South Central Texas data on lizards and turtles. Another consideration for preparation of this book was that it would complete the reptiles of the region when coupled with *Snakes of South Central Texas* by Thomas G. Vermersch and Robert E. Kuntz, which was published in 1986.

Lizards and Turtles of South Central Texas brings before a popular audience of serious students, natural history enthusiasts, farmers, ranchers, landowners, and people who engage in outdoor recreational activities, a portion of the rich biotic diversity present in South Central Texas. As residents of the region steadily become more aware of the concepts of conservation, protection, and coexistence of the nongame wildlife, let us hope that ordinary citizens, community leaders, and developers comprehend how our present actions will impact the biological diversity. Some of our traditional policies in South Central Texas threaten to eradicate several wildlife species.

The primary focus of the text is devoted to the recognition and biology of twenty-two species and subspecies of lizards, ten species of turtles, and one specie of tortoise that inhabit South Central Texas.

Eight counties encompassing 7,320 square miles have been designated as "South Central Texas" for the purpose of this work. These counties include Atascosa, Bandera, Bexar, Comal, Guadalupe, Kendall, Medina, and Wilson.

The glossary is provided for individuals at different grade levels who may have difficulty in understanding the terminology. Drawings of external characteristics (scalation) used in lizard and turtle identification are found in the introductory section. This section can be used in conjunction with the identification sections of each write-up for familiarization with scalation characteristics, e.g., counting femoral pores only on one side, etc. Additionally, an effort has been made to simplify some of the technical terminology used by herpetologists.

To further simplify primary diagnostic features in the identification section, I have italicized the primary identification features. Color photographs are provided in one section to supplement general identification on thirty-two species and subspecies covered in the text (no color photograph could be obtained for the southern earless lizard, *Holbrookia lacerata subcaudalis*). Due to the number of publications involving the South Central Texas herpetofauna, I have found it impossible to individually cite all researchers and their work. However, if these are not cited in the text, they are included in the bibliography. This sizable bibliography should help students who wish to inquire further into the biology of these interesting organisms.

A state and regional map, which indicates the distribution, is provided on each turtle and lizard. These maps are based upon accumulated data from various sources, including Raun and Gehlbach (1972), Conant (1975), Dixon (1987), Garrett and Barker (1987), and new county records compiled from universities and museums by James Dixon, C. T. McAllister, Ralph Axtell, Carl Lieb, and others. It should be noted that the original data collected concerning Raun and Gehlbach's 1972 state range maps have been lost. According to Raun (pers. comm., 1989), these data included several observational records by colleagues, the origin of which is now difficult to substantiate. I have included only a few personal observations and collected specimens from reputable amateur herpetologists and professionals who have directly observed and identified specimens firsthand in the field. Occasionally, this type of information is more useful than some museum specimen, when the collector is no longer living or cannot be found to answer a question concerning its collection. Most of the data on distribution are indicated by dots on the enlarged county maps which accompany state maps of distribution. An asterisk (*) marked on a county map indicates a literature record without exact locality data.

These distribution data have been further supplemented by my examination of miscellaneous records and fixed specimens in museums and schools throughout the region. The data base, consisting of 1,700 records on museum and observational accounts of lizards and turtles of South Central Texas, is available from Living Jungle Science Programs, San Antonio, Texas.

The text is organized to feature important categories of information. These categories include:

Introduction — General information on total distribution and interesting features are highlighted.

Identification — General diagnostic description of color, pattern, scalation, and other external features (adult and juvenile descriptions included). These are contrasted with other species of similar forms. Significant diagnostic features are italicized in the identification section. It should be noted that some of these diagnostic features for identification are based on published literature (listed in the bibliography) and may not encompass local variation in color, pattern, scalation, and other external features.

Size — External measurements to understand the total normal and maximum lengths in adults. Measurements (U.S. and metric equivalents) taken on lizards are measured from the tip on the rostral scale to the tip of the tail. Lizard snout-vent lengths are not recorded because of the lack of concise data on all forms recorded from the region. Turtles are measured from the front of the nuchal scute (in front of the carapace) directly back to the tip of the (posterior) supracaudals.

Behavior — Behavioral features may include documented activity periods, home range, territory, social relationships, aggregations, courtship, shelter, environmental and physiological side effects, and other behavioral traits. Temperatures affecting the behavior are often rounded off from Fahrenheit to Centigrade equivalents.

Food — Dietary considerations may include primary and secondary preferences, occasional examination of stomach contents, natural and captive diets, and seasonal or other shifts.

Habitat — Major types and associations of vegetation, soil, water, and other factors that comprise the species' habitat preference and overall distribution in South Central Texas are described.

Breeding — Reproductive data may include oviparous or viviparous information, mating season, nesting, or egg-laying season, fertilization information, brooding, incubation time and temperatures, communal egg-laying, egg deposition in nests, clutch size and frequency, egg measurements, hatchling measurements (see size section above), juvenile growth and development, and other pertinent reproductive information documented on the species or subspecies. Presentation of reproductive data given in this book may be based on area (regional, state) or on all species data, depending on what current information is available. Thus, species-level information is occasionally used where regional subspecific data is lacking. It should be recognized, however, that several factors may influence local variation in reproductive biology of reptiles. For example, variation in sex and age influences clutch and litter size, with larger and older females normally producing larger broods. Moreover, availability of food, length of growing season, yearly fluctuations in weather, and other environmental factors have important effects on reproduction.

General Information — This section includes a wide variety of available information, including relative abundance, longevity, predation, economic importance, references to type localities, and recent research contributions.

In addition to the bibliographic materials cited, other useful sources of information for this and the other sections in this book have included the South/Central Texas Herpetological Society, the Texas Herpetological Society, the Society for the Study of Amphibians and Reptiles, the American Society of Ichthyologists and Herpetologists, the Herpetologists' League, the San Antonio Zoo Reptile Department, the Witte Museum Natural History Department, and the academic research centers in herpetology at Texas A&M University, Baylor University, and the Universities of Texas at Arlington, Austin, El Paso, and Tyler.

(Information compiled as of 1990.)

Acknowledgments

This work would not have been possible without reference to the accomplishments of others or suggestions and criticisms of colleagues, students, and friends. I am indebted, directly or indirectly, to a number of biologists and herpetologists for their assistance.

I am particularly pleased to acknowledge the assistance of Carl S. Lieb for his valuable contributions, which include part of the introduction section and comments regarding the final manuscript, and my wife, Carol Vermersch, for her assistance and understanding in proofreading and typing the drafts of yet another wildlife manuscript for me.

Special thanks are extended to Ralph W. Axtell, Richard J. Baldauf, Bryce C. Brown, Stephen Cheyney, Joseph T. Collins, W. Rowe Elliott III, Carl H. Ernst, Gary W. Ferguson, John B. Iverson, R. Earl Olson, Eric R. Pianka, Andrew H. Price, Hobart M. Smith, and Stan E. Trauth for thoroughly reviewing and commenting on the content of the text.

Additional appreciation is extended to David Barker, Mike Bishop, Ray D. Burkett, Allen Chaney, Trent Cheyney, Mike Dawson, Bruce Deuley, Ed Farmer, Jerry Fischer, Neil Ford, Joe Forks, Randy Glickman, Jim Godwin, David Gyure, David Haynes, Terry Hibbitts, John Hollister, Frank Judd, Sara Kerr, F. Wayne King, Carol Koop, Robert Kuntz, Lionel Landry, E. A. Liner, David Lintz, Jon Lowell, Roy McDiarmid, D. Craig McIntyre, John McLain, Greg Mengden, William Montgomery, Holly Morgan, Robert Norris, Tony Sarratt, Diana Sarratt, Niels Saustrup, Robert Schattel, A. J. Seippel, Rod Towers, Wayne Van Devender, Rick Van Dyke, Ben Wayment, and John Wright for their information and overall support.

David Barker, Terry Hibbitts, Robert Kuntz, W. B. Love, and R. W. Van Devender provided color photographs of some of the lizards and turtles. I am also especially grateful to Allison Smith for her line drawings of the external characteristics of the lizards and turtles in the introduction section, and to James R. Dixon for providing county record information. Some of the lizard and turtle illustrations in the external characteristics (scalation) section are redrawn from Dennis and Barlowe (in Smith and Brodie, 1982).

Introduction

Sixty-five million years ago the Mesozoic Era came to an end, and with it the time of the dinosaurs and other spectacular forms of reptilian life. The secret of the demise of these strange and wonderful reptiles is a mystery that may never be satisfactorily unraveled, but it is clear that they are indeed gone, and that other forms of vertebrate life have evolved to fill niches the ancient reptiles vacated. It is now we mammals that fill the earth in such familiar profusion, and it is hard to visualize that Mesozoic world where mammals were comparatively few and seemingly as unimportant as the modern reptiles are today.

The legacy left us by the dinosaurs are the crocodiles and birds, their only true descendants. The other living reptile groups include the turtles, the lizards and snakes, and a lizard-like reptile called the tuatara. Turtles are true survivors of the time of the dinosaurs: those that swam in Mesozoic seas are the same unmistakable shelled reptiles that live in modern waters. Lizards also first appeared in the Mesozoic, but as their major evolution took place mostly against the backdrop of the mammalian radiations of the Cenozoic, they seem relative newcomers. Snakes evolved out of a group of lizards and are thus younger still. The tuatara, restricted now to the New Zealand area, is of uncertain relationship. It was once thought to also be a surviving remnant of a Mesozoic lineage, but there is now recent controversy as to its possible lizard affinities.

The book you hold in your hand concerns the modern reptile groups, exclusive of the snakes, that are characteristic of the South Central Texas area. The snakes have been treated by an earlier book (Vermersch and Kuntz, 1986), and the crocodilian representative, the American alligator, has also been excluded. Alligators are historically known from the South Central Texas region and have persisted in certain South Texas rivers and cattle tanks. Several decades of vigorous

1

federal and state conservation efforts have resulted in substantial population increases along the Texas coast, and it is not unlikely that inland dispersal may someday restore alligators as an obvious element of the regional herpetofauna.

Reptiles as a group share several attributes that separate them from other terrestrial vertebrate (backboned) animals. Reptiles and amphibians are placed together within the scientific discipline of herpetology mostly as a matter of convenience; they are actually very dissimilar in their evolutionary history, structure, and physiology. Clearly, reptiles are better adapted than amphibians for life in dry environments, as their comparatively waterproof skins demonstrate. However, and more important to the overall comparison of amphibian versus reptile evolutionary success, the reptilian egg possesses an *amnion* within it, whereas the amphibian egg does not. The amnion is a membrane that prevents the embryo from drying out when the egg is laid in a relatively dry terrestrial situation. In contrast, the eggs of amphibians must be laid in water or in some damp environment on land.

Differences between living groups of reptiles and mammals include several anatomical features, including in mammals the presence of hair, mammary glands, three middle ear bones, and the pattern blood-vascular circulation. In addition, both birds and mammals are *endothermic,* meaning that the appropriate range of body temperature needed for physiological function is maintained by heat-producing chemical reactions within the animal. On the other hand, modern reptiles are *ectothermic,* relying on behavior patterns and an external source of heat (the sun) to maintain their body temperature within an effective physiological range. These terms replace the outmoded and somewhat misleading expressions "cold-blooded" and "poikilothermic." A turtle basking on a log on a hot summer day can be very "warm-blooded" indeed, and a lizard darting between hot sunshine and cool shade while foraging is regulating its body temperature with considerable precision.

The ectothermic attributes of modern reptiles are occasionally invoked as an inferior physiological trait to the endothermy seen in mammals and birds. However, ectothermy is better viewed as simply different, with strong advantages. A reptile slows its metabolism with cool environmental conditions, unlike its mammalian and avian counterparts that must continually expend energy simply to keep warm. Such expenditure of energy requires additional nutrition, and thus the possibility of starvation always lies in the immediate future for an endothermic animal. It is not surprising that such adaptations as hibernation and torpor in endothermic animals have evolved; these essentially represent an attempt to temporarily escape the nutritive demands of endothermy by utilizing reptilian energy-conservation strategies.

It is unlikely that anyone could mistake a turtle for any other living animal, but nevertheless a confusing roster of names is applied to them: turtle, tortoise, terrapin, and slider are common appellations in Texas. The expression "turtle" is the most inclusive term and can apply to any member of the reptilian order Testudines. Tortoises are terrestrial turtles, usually restricted to members of the family Testudinidae (such as the Texas tortoise, *Gopherus berlandieri*). "Terrapin" is correctly applied only to the diamondback terrapin of the Gulf Coast, but the name is often used for the box turtles (*Terrapene*) of the family Emydidae. To make matters worse, the terrestrial box turtles sometimes are also called "tortoises." And finally, certain aquatic members of the family Emydidae are called "sliders," presumably because they slide off logs into the water as soon as one comes upon them.

By whatever name they are called, the persistence and success of the turtles from the time of the dinosaurs to today is based on what has been for them a particularly successful adaptation: enclosure of the body in a protective bony shell. The bones of the backbone exclusive of the neck and tail are fused to the upper part of the shell (carapace), and the important internal organs are packed into the shell cavity. The skull tends to be heavy and massive, and a bony beak (instead of teeth) is present. Such a body plan has limitations for the exploitation of certain types of ecological niches; there are no arboreal or flying species. But within the limits imposed by life in a bony box, turtles are conspicuous over such parts of the world that are warm enough for ectothermic animals to live. Moreover, some modifications to the size and degree of hardness to the shell are seen in different groups: the lower shell (plastron) of the snapping turtle is greatly reduced, and a significant portion of the shell of soft-shelled turtles consists of cartilage instead of bone.

All turtles are egg-layers, the eggs being buried on land in a simple nest. Once the eggs are deposited and covered, the female parent abandons the nest. Whatever eggs and hatchlings survive the numerous predators (including humans) are on their own. Some lizards and snakes have many species where the eggs are retained in the body of the mother until they are about to hatch, and parental care occurs in the crocodilians (and probably occurred in many dinosaurs). Another interesting feature concerning turtle eggs is their temperature-influenced sex development: the eggs of several species will produce a predominance of either males or females depending upon the prevailing incubation temperature.

Turtles are famous in story and fable for their long lives. These are mostly exaggerated, although several species of turtles are reported to have lived four decades or so in captivity, with the large tortoises capable of carrying on for at least 150 years. Existence over several centuries, however, is undoubtedly a myth. This myth continues to be perpetuated by pranksters who carve impossible dates into the shells of wild turtles before releasing them.

Other than their status as a mythical symbol of longevity, the principal role of turtles in human society is that of food item. The large forms, particularly the sea turtles and tortoises, have been important food resources in historical times; excessive consumption as such has contributed significantly to their decline in abundance. Two other popular edible species, both of which occur out of the South Central Texas area, are the diamondback terrapin, of the Atlantic and Gulf Coastal areas, and the alligator snapping turtle, found in large rivers in the Southeast. As harvesting of these species from the wild takes place in the absence of significant wildlife management, it is not surprising that they are also in apparent decline over much of their ranges. Lastly, several species of slider and soft-shelled turtles are consumed as part of the local cuisine, particularly in parts of the southern United States where these aquatic forms are abundant and easily harvested.

The other principal relationship between men and turtles is that of keeper and pet. Whereas the adults of many species make satisfactory captives that require relatively little care, some forms, and the hatchlings of most, are best left to experienced reptilian husbandrymen. Some aquatic species of turtles also can interact with humans through the transmission of disease brought about by *Salmonella* bacteria and related microorganisms. The potential for such infection can be greatly reduced through normal hygienic precautions, including hand-washing after picking up or otherwise handling a turtle.

In contrast to turtles, the popular conception of lizards is that of a darting quickness in the sunshine, a transient presence in the garden, or a fleeting glimpse crossing the highway. This concept for these active, successful reptiles has largely replaced the older notion of languid laziness; it is not surprising that the term "lounge-lizard" has largely disappeared from the average American's vocabulary.

4

Most lizards are insectivorous, have four legs and a relatively long tail, and lead short lives. Only a few large species are herbivores or non-insect eating carnivores; these forms occur well outside of the South Central Texas area. The snake-like legless lizard (*Ophisaurus attenuatus*) is historically known from this area but is now quite rare or even extirpated. The few local salamanders also have four legs and a tail, but these amphibians are easily distinguished from lizards by their lack of scales and by the absence of claws on the ends of the digits. Most small lizard species probably live in the wild only one or two years, whereas some of the larger species have survived in captivity up to a dozen or more years. Most of the lizards in South Central Texas lay eggs; the females of some, such as the skinks, remain with the eggs for some time after they are laid.

A variety of terrestrial habitat types are occupied by lizards, and frequently certain species are associated with a specific habitat. For example, one typically sees Texas spiny lizards on the trunks of trees, and prairie-lined racerunners in open areas with fine sandy soil. Whereas many lizards are conspicuous, a few are quite secretive and rarely seen; moreover, the two local species of geckos are entirely nocturnal.

Lizards are the favored prey of several species of local snakes, as well as some carnivorous mammals and a few birds. In all of the local species (except the collared lizard), the tail is capable of breaking off when grasped by a predator, thus allowing the lizard to escape. The broken tail regenerates in part, being replaced by a dark brown or gray cartilaginous structure.

The importance of lizards to humans is mostly through their insect-eating habits, which make them valuable additions to most human habitations and gardens. Unfortunately, because lizards and other insectivores depend primarily upon insects as a food resource, they are potentially vulnerable to pesticide misuse, both through the diminishing of their potential food supply and through the concentration of toxic residues in their body tissues caused by ingestion of contaminated insects.

Only the very largest lizard species are satisfactorily utilized in human commerce for food and hides, as part of the (largely unmanaged) worldwide wildlife exploitation industry. As far as is known, lizards have not been directly implicated in the transmission of diseases to humans, although there is some circumstantial evidence that they (as well as many other animals) may act as reservoirs for some forms of mosquito-transmitted encephalitis.

5

There are no naturally occurring venomous lizards in South Central Texas, in spite of the folklore that the alligator lizard ("Escorpion") is dangerously poisonous. Nevertheless this lizard, as well as other large species, can deliver a painful pinching bite when handled carelessly.

The rich biotic heritage of the South Central Texas area is apparent. The tall-grass prairies of central North America, the mesic woodlands and riparian communities of the east, the subtropical brushlands of the Tamaulipan plain, and the Balconian foothill extension of the Sierra Madres all converge in this region to meet and mix their biotas. The patterns of distribution seen in the reptiles of the South Central Texas area also reflect these biotic assemblages, giving the region an exceedingly rich and diverse herpetofauna.

Unfortunately, the South Central Texas area habitats for reptiles and other wildlife, like those of much of the southwestern United States, are undergoing progressive deterioration. This decline is correlated with a variety of factors but is clearly tied to poor land use practice and burgeoning human populations. Not much can be done about the latter; in the eight-county region covered by this book, human populations have by now exceeded one and a half million. Most are concentrated in the San Antonio megalopolis, which was probably the area of greatest plant and animal diversity in the entire state only a century ago. As for land use, particularly pressing is the need for protection and conservation (not necessarily "development") of water resources, and the safeguarding of surface and ground water from overexploitation and pollution. Additionally, more caution needs to be exercised in the use of insecticides, herbicides, and the disposal of toxic chemicals; soil conservation practices, not only those involving ranching, farming and rural life but also those associated with urban development, still need enhancing. Ominously, the long-term planning and foresight needed to address the conservation of our natural heritage have always taken a back seat to short-term political and economic expediency. This approach will have to change, and soon, if there is going to be any biotic diversity left for future generations to enjoy.

Conservation of most wild organisms begins with protection of their habitats, not necessarily with direct protection of individuals. In recent years, however, it has become alarmingly apparent that local or even statewide efforts are not enough, and that widespread (and perhaps global) threats are beginning to show their effects upon taxa. The sensitivity of reptiles to these threats will vary, with some species ap-

6

parently more sensitive than others to environmental changes, pollution, and epidemic disease. The sudden disappearance or decline of formerly abundant forms, such as horned lizards or map turtles, should serve as an "early warning" to be ignored at our own peril.

— CARL S. LIEB

Laboratory for Environmental Biology
University of Texas at El Paso

Contributions to the Knowledge of Lizards and Turtles of South Central Texas

Numerous ethnic groups have contributed to the rich history of this region. Reptiles such as lizards and turtles have found their way into the folklore, art, and even the diet of early South Texas inhabitants. Indian cave art of the Trans-Pecos area to the west includes clearly depicted head and body configurations which represent lizards and turtles found in Texas. However, these pictographs occur in varied contexts, making accurate interpretations difficult.

From the early 1500s to the late 1600s, Spanish explorers traveling across Texas (including a number who visited South Central Texas) reported various reptiles. The first significant account on the culture and biology of the region is found in the diaries of Alvar Nuñez Cabeza de Vaca. According to Newcomb (1961) quoting from these diaries (circa 1533), the natives (Yguaces and Mariames, generally referred to as the "Coahuiltecans") used a wide variety of living creatures as food. Specifically mentioned in their list of food items were spiders, ant eggs, worms, lizards, and snakes.

By the eighteenth century, Europeans established themselves in South Texas, especially in response to the development of the missions, e.g., Mission San Jose in San Antonio. The missionaries of this time were primarily interested in developing agriculture and saving souls. However, many were also of a scholarly bent, and general conditions of climate, native vegetation, and some wildlife information were documented during this period.

Jean Louis Berlandier, a French botanist who resided in Mexico, was one of the first scientists to document his travels from San Antonio de Bexar to various locations around the state from 1828 to 1834. On his first trip from San Antonio de Bexar to Eagle Pass and Laredo (June 10 to July 28, 1834) he encountered a Texas tortoise, and as he crossed the Nueces River he came upon soft-shelled turtles. Berlandier remarked on how common the soft-shelled turtles were around San Antonio de Bexar and noted that on a return trip from San Felipe to San Antonio de Bexar he saw box turtles. Also in 1829 was his first encounter with a Texas horned lizard (Ohlendorf, Bigelow, and Standifer, 1980).

Following the Mexican War (1846–1848), the United States government sponsored several boundary surveying expeditions in the Southwest. These often passed through San Antonio, at the time a major economic and military center for the fledgling state of Texas. During these surveys (1850 through 1855) several military men collected reptiles or otherwise contributed to what marks the beginning of Texas zoology. These men included John R. Bartlett (the first U.S.-Mexican boundary commissioner), Col. J. D. Graham, Maj. John H. Clark (his assistant who served as zoologist), Maj. William H. Emory (who took over the direction of the boundary survey), Capt. Randolph B. Marcy, Capt. Arthur C. V. Schott, Capt. Van Vliet, and Army Surgeons Kennerly and Woodhouse. Their collections were usually sent to the Smithsonian Institution to be studied by Professors Spencer Baird and Charles Girard.

Subsequent to this pioneering work, herpetological activity in the South Central Texas area began to accelerate. C. W. Schuermann in 1878 sent a collection of reptiles from San Antonio to the Smithsonian Institution. Edward D. Cope, the outstanding American herpetologist of the nineteenth century, collected in 1886 along the rocky hills northwest of San Antonio. Cope was affiliated with the Academy of Natural Sciences of Philadelphia. His collecting location was at or near the Helotes Creek homestead of Gabriel Marnock. Marnock moved to the Helotes area around 1889, as he was fascinated by this almost virgin territory and its numerous species of amphibians and reptiles. He spent more than forty-two years (the remainder of his life) collecting reptiles and amphibians. Most of these specimens were sent to Cope or to John K. Strecker at Baylor University. Another significant collection of that era was made by Louis Garni, a self-taught naturalist who collected in San Antonio in Bexar County and Boerne in Kendall County.

Even though John Strecker, in his 1933 paper on collecting at Helotes, called Gabriel Marnock a "pioneer Texas herpetologist," Strecker himself is usually considered the father of Texas herpetology. By 1915 he had published the first definitive list of 163 species of Texas reptiles and amphibians. Moreover, Strecker's influence as a herpetologist grew as he accumulated at Baylor University the largest collection of Texas reptiles and amphibians anywhere, and over the span of his career wrote sixty papers on the reptiles and amphibians of Texas.

A second important figure in the development of herpetological knowledge in the South Central Texas area was A. J. Kirn of Somerset. Kirn was a naturalist active in the region from the early 1920s to 1950, and, whereas he did not publish many papers on herpetological subjects, he carried on significant correspondence with several of the leading zoologists of the time. Some of his notes were eventually published by Cornell herpetologists Albert H. and Anna A. Wright in their classic books on frogs and snakes of the United States. The Wrights met Kirn in 1925 through their mutual friends, Roy and Ellen Quillan of San Antonio. The Quillans, noted field naturalists themselves, also advised the Wrights on regional herpetological subjects.

From the 1950s through the 1970s, although many herpetologists visited the South Central Texas area on collecting trips, few remained for any length of time in residence. Exceptions included Ralph Axtell, now a premier authority on Texas lizards, and R. Earl Olson, formerly of the Witte Memorial Museum. Carl Lieb, San Antonio native and field companion of the author as a youth, kept detailed records and observations on the local herpetofauna during the 1960s; he has since continued to study some of its components from afar. More recently, dozens of herpetologists, biologists, and naturalists have additionally contributed to the understanding of the South Central Texas herpetofauna. As many of these were contacted during the preparation of this book, the reader is referred to the acknowledgments for a partial listing.

EXTERNAL CHARACTERISTICS (SCALATION) USED IN LIZARD IDENTIFICATION

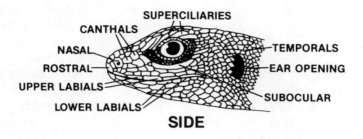

SIDE

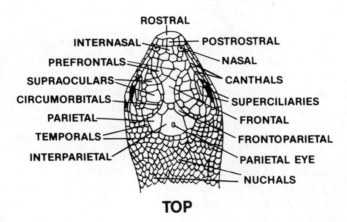

TOP

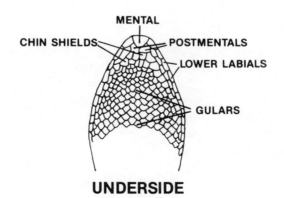

UNDERSIDE

UNDERSIDE OF TOE

TOE LAMELLA   SMOOTH   KEELED

COUNTING DORSAL SCALES
AND TOTAL MEASUREMENT

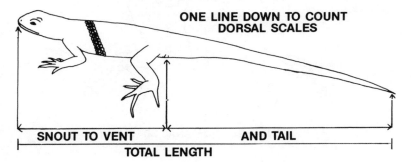

ONE LINE DOWN TO COUNT
DORSAL SCALES

SNOUT TO VENT   AND TAIL

TOTAL LENGTH

DORSAL
KEELED SCALES

DORSAL
GRANULAR SCALES

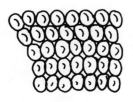

SMOOTH ROUNDED
SCALES

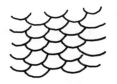

QUADRANGULAR
VENTRAL SCALES

11

POSTERIOR
UNDERSIDE

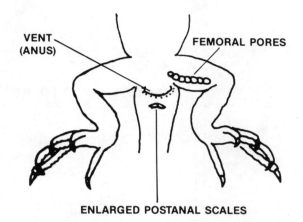

VENT
(ANUS)

FEMORAL PORES

ENLARGED POSTANAL SCALES

POSTERIOR
UNDERSIDE

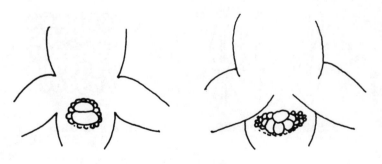

SINGLE LARGE PREANAL

SEVERAL PREANALS

EXTERNAL CHARACTERISTICS (SCALATION) USED IN TURTLE IDENTIFICATION

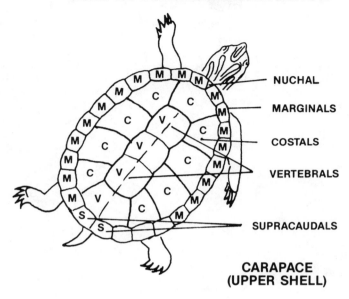

NUCHAL

MARGINALS

COSTALS

VERTEBRALS

SUPRACAUDALS

CARAPACE
(UPPER SHELL)

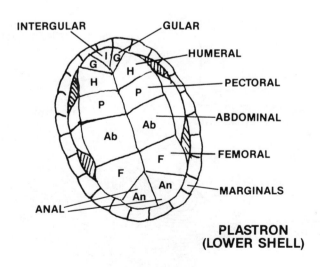

INTERGULAR   GULAR

HUMERAL

PECTORAL

ABDOMINAL

FEMORAL

MARGINALS

ANAL

PLASTRON
(LOWER SHELL)

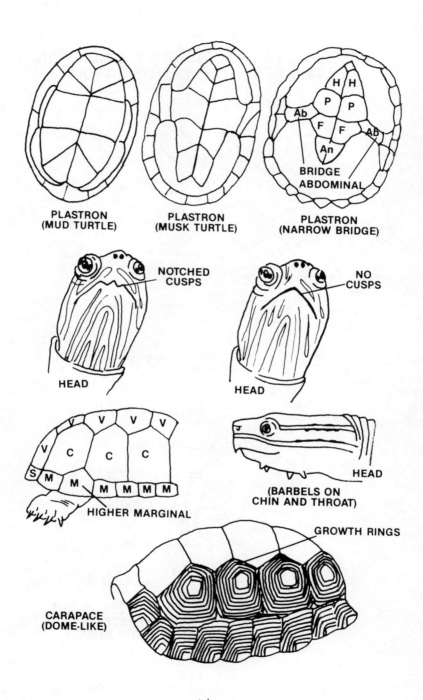

PLASTRON
(MUD TURTLE)

PLASTRON
(MUSK TURTLE)

PLASTRON
(NARROW BRIDGE)

H H
P P
Ab F F Ab
An
BRIDGE
ABDOMINAL

NOTCHED
CUSPS

HEAD

NO
CUSPS

HEAD

V V V V
V C C C
S M M M M M M
HIGHER MARGINAL

HEAD
(BARBELS ON
CHIN AND THROAT)

GROWTH RINGS

CARAPACE
(DOME-LIKE)

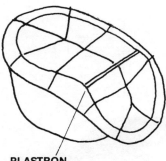

**SOFTSHELL CARAPACE
(TUBERCLES SCATTERED)**

**PLASTRON
(HINGED)**

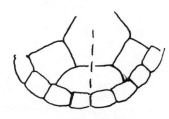

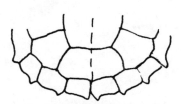

**CARAPACE
(SMOOTH)**

**CARAPACE
(NOTCHED)**

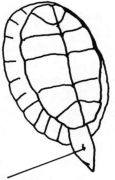

**SHELL
(MALE REAR VIEW)**

**(ANUS BEYOND SHELL
ON MALES ONLY)**

Geckos
(Family — Gekkonidae)

TEXAS BANDED GECKO

Coleonyx brevis Stejneger 1893

This nocturnal, terrestrial lizard has relatively short legs, a flat body, and large eyes. It is the only native South Central Texas member of the family Gekkonidae, a large, worldwide group of some 85 genera and about 650 species. Three species of *Coleonyx* are found in the United States, with the Texas banded gecko occurring in the northern Chihuahuan Desert of New Mexico, Texas, and Mexico, as well as in parts of the Rio Grande Plain in the latter two areas. In the South Central Texas area this species is documented only from the southwestern margin of the Edwards Plateau.

Identification: Dark to light brownish bands, wider than the light yellow to tan-colored narrower bands between them, are found across the back. These bands break up and fade before reaching the ventral border area. Large adults, as they increase in age, usually have their bands interrupted by a mottling of dark brown markings, usually in the form of spots and blotches. Juveniles have distinct, dark brown crossbands that alternate with slightly smaller ones that are a pale yellow or pink to cream ground color.

The head is slightly larger than the body, and the large eyes are outlined by a cream or pale yellow color. Other features include vertical pupils, *movable eyelids, no toe pads on the slender toes,* three to six preanal pores in the males (in two series separated by one or more median scales), with a flat lateral spur on each side of the tail behind the hind limbs. This gecko has fine, uniform, granular scales covering its body. The thick but fragile tail is about the same length as the body. Tail regeneration following tail loss is a slow process, and normally the new tail will not reach its former length. No similar-appearing species are found in the area.

Size: Adults range in size from just under 4 inches to $4^7/8$ inches (10.2 to 12.4 cm) in length.

Behavior: This nocturnal gecko may live for three or more years, according to Axtell (pers. comm.). During the day it is found sleeping under flat rocks, usually becoming active just after sunset. When this lizard is first picked up, or if bitten by another gecko when fighting, it may emit a faint squeak. Other behavioral characteristics include a defensive tail display and male aggressive behavior. The defensive tail display in adult lizards apparently facilitates escape from small snakes and other predators. This is achieved by elevating the tail well above the ground and undulating it slowly from side to side. Something similar is also accomplished when the gecko is fleeing with its tail coiled vertically to one side of its back. Three out of four of these lizards have regenerated tails, indicating that the detachable tail is one of the primary escape mechanisms for this lizard (Dial, 1978). Tail-twitching is often observed when the gecko stalks its prey. Geckos have excellent vision and are able to seize a moving insect in a flash.

Male aggressive behavior involves the male rising up high on all four legs and arching its back with its head held in a low position and the throat inflated. If the other male does not flee, it is stalked and nudged on the body and tail. Both lizards may do this for a short time until one lashes out at the rival male and bites it. This ritual may be repeated several times during antagonistic male territorial defense. Although these confrontations are brief and usually harmless, territorial dominance is established by one male.

The thermal ecology of this lizard suggests that, when under rock shelters in their natural habitat, they may behaviorally thermoregulate. Laboratory observations suggest that on cool, sunny days, when the sun's radiation increases the rock temperatures above the air and substrate temperatures, they may elevate their bodies to contact the warmer rock above.

Food: The diet includes soft-bodied insects (especially termites), beetles, small terrestrial arachnids, and other available arthropods that they may find while prowling at night. Like many other geckos, they will eat their own shed skin when it has molted.

Habitat: In South Central Texas they are found along the southern edge of the Edwards Plateau. There they occur along rocky limestone slopes and outcrops along canyons, taking shelter during the daytime under flat rocks, in rock crevices, and rarely, under logs and discarded rubbish. They may descend into holes or deep crevices as the ground dries and the weather becomes hotter. These lizards are locally common in certain areas around Medina Lake and were historically abundant in the Helotes region. In the Chihuahuan Desert proper, the geckos are commonly found in rocky outcrops along floodplains, on rocky slopes, and in arid canyons in association with cedar-ocotillo and the rimrocks of the persimmon-shinoak. It is not unusual to spot these fat-tailed, little creatures walking across the pavement on desert roadways at night.

Breeding: Texas banded gecko mating has been reported in spring and summer in captivity. These lizards may attain rapid sexual maturity, and two or three clutches of smooth, white eggs are produced per season. The clutch size, however, is constant at two eggs per clutch. The newborn lizards are approximately 1³/₄ inches (4.5 cm) at hatching.

General information: Edward D. Cope, an outstanding herpetologist of the nineteenth century, wrote in 1886 that he had found this lizard rather abundant in the rocky hills of the first plateau northwest of San Antonio, but he did not observe it in the region north of the Guadalupe or Llano rivers. According to Axtell (1986), variations in density appear to be irregularly cyclical with high density in one location some years and very low during others. Although data is lacking, possibly this is related to short-term climatic changes. The type locality for *Coleonyx brevis,* the place from which the original specimens were described, is Helotes, Texas, in Bexar County. This specimen was collected in 1893 by Gabriel Marnock.

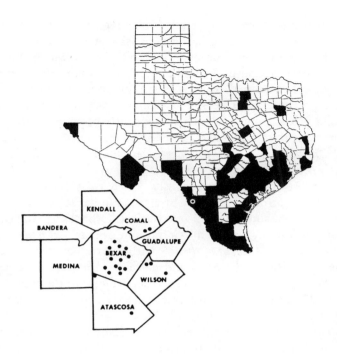

MEDITERRANEAN GECKO

Hemidactylus turcicus (Linnaeus) 1758

    This Old World gekkonid lizard is naturally distributed from western India to Somaliland, along either side of the coastal Mediterranean Sea from Egypt to Spain, Morocco, and the Canary Islands. It has been introduced along the Gulf Coast of the United States into mainland Mexico and has established populations in Panama, Puerto Rico, and Cuba. One of the first reported Mediterranean geckos was found in Texas in 1933 along the Gulf Coast (but no locality was given) and subsequent specimens were reported from Brownsville and at the old Market Square produce terminal in San Antonio around the end of World War II. It was part of a small established population in the downtown area of San Antonio, and may have been introduced there from the large Brownsville colony. Populations are now found in El Paso, San Antonio, Austin, Fort Worth, Waco, Houston, Galves-

19

ton, Corpus Christi, the Rio Grande Valley region, the Big Bend, and many other Texas areas, and are usually associated with human habitation. It is fascinating to observe the rapid pace at which this introduced species is expanding its range throughout the southeastern Gulf Coast states. Few animals rival it in its colonization potential. Their success is probably due chiefly to communal nesting in building materials and produce crates, which subsequently provide a moving van-like service for the lizard and its eggs.

Identification: This semitranslucent, pinkish to white or pale gray lizard has small, irregular, dark brown to grayish spots on the entire dorsal surface. It has 14 to 16 rows of conspicuously keeled *warty tubercles* along the dorsal surface, surrounded mostly by tiny, granular scales. The body is usually flattened, and the pupils are vertical. Other external characteristics include large eyes, which lack movable eyelids, *toes with wide pads that extend almost the length of the toe,* with a terminal claw. The male possesses 2 to 10 preanal pores and no femoral pores. Juveniles have banded tails. No similar-looking species are found in the area.

Size: The average head-to-tail length for an adult lizard is approximately 4 to 5 inches (10.2 to 12.7 cm).

Behavior: This nocturnal gecko is especially partial to the walls of dwellings that have outdoor lighting which attracts insects at night. The vision of these lizards plays a dominant role in their foraging for food. This species of lizard has very few natural predators and few interspecific competitors. It is highly territorial and will defend its favorite foraging spot, such as a screened window near a night-light that attracts insects. However, communal nests are often used by several individuals. A faint squeaking sound or bird-like chirping may be emitted during territorial fighting or other stress-related behavior. When threatened, they may escape into cracks of bricks and old, wooden window frames. These lizards are most active in the first several hours after nightfall, gradually tapering off their activity after midnight.

Hemidactylus turcicus is capable of climbing up and down smooth or rough walls or crossing a ceiling with its specially expanded toes with comb-like pads (lamellae) and claws. Each of the comb-like pads has thousands of soft, tiny, hair-like appendages that hold the fast-moving lizard to the smooth wall or ceiling (Hildebrand, 1982). Mediterranean geckos, at one site, moved an average of 18 feet (6m) between their initial site and recapture site, according to Rose and Barbour (1968). A more recent study by Selcer (1986) indicates density estimates ranging from 544 to 2,210 lizards per hectare (2½ acres), depending on the method, year, and study area. He also reported that the geckos' movements between metal buildings were infrequent. Even the most severe winters in South Central Texas have had only a minimal effect on the population size in the author's household-storeroom gecko colony. Surely this species must benefit from human heat production in overwintering.

This lizard cleans or moistens its eyes by licking across them with its flat tongue, and is capable of detaching its tail if grabbed. Tail regeneration was almost complete in 35 days for one documented juvenile, but the time was slightly longer for the adult (taking 60 days or more to regenerate its tail).

Food: These geckos have been observed eating caterpillars, moths, ants, small beetles, cockroaches (especially hatchling American cockroaches), earwigs, homopterans, and even mosquitoes. Virtually any flying insect attracted to a light at night is subject to being snapped up by a gecko should it land near it on a wall or window screen.

Habitat: Man-made habitats are frequented and may include under stacked boards, baseboards of walls, and window ledges, behind storm drains, in rock dwellings that are illuminated with a bright light, and around homes, warehouses, storerooms, garages, lumberyards, motels, truck stops, and other public buildings in association with human habitation. In residential neighborhoods it is commonly found inside houses where, if tolerated, it thrives in a niche similar to that of its close relative, the aptly named Asian house gecko.

Breeding: The mating season for reproductively mature lizards usually lasts from early April through late July (Selcer, 1986). Copulation has been observed in June and July. The first females observed with large eggs (which can be seen as white objects through the semitranslucent skin of the abdomen) appear during late May, and the last females with eggs were observed in early August. These late eggs should hatch toward the end of September, approximately 40 days after being laid. Usually 2 eggs are laid per clutch with usually 2 (or possibly 3) clutches per season, according to Rose and Barbour (1968) and Selcer (1986). The eggs are hard-shelled, calcareous, and average $3/8$ inch (10 mm) by $7/16$ inch (11 mm). Such hard-shelled eggs will not desiccate as rapidly as parchment-shelled eggs (Dunson, 1982) in other lizard species. Rose and Barbour (1968) reported nest building (debris placed to cover the eggs), and the author has observed these lizards laying eggs in the empty tooth pockets of the lower jaw of an alligator skull on his storeroom shelf, as well as in other areas of the room.

Hatchlings are about 1³/₄ to 1⁷/₈ inches (4.5 to 4.7 cm) in total length. Selcer (1982) reported that the age at maturity of 20 known-age lizards ranged from 5.9 to 10.9 months.

General information: Observations of this gecko have been made on a year-round basis in South Central Texas, including several observations during January and February in the author's sheet metal storeroom. This storeroom would heat up like a greenhouse on sunny days during the winter, but at least one gecko has been reported active on a building at an air temperature of only 42 F. (5.5 C.). This lizard has a relatively long life expectancy, commonly living for three years or longer (according to author's records and reconfirmed by Selcer, 1986).

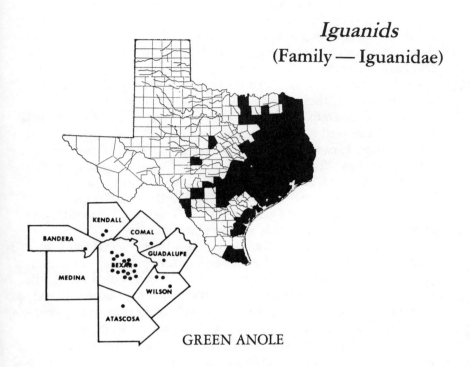

Iguanids
(Family — Iguanidae)

GREEN ANOLE

Anolis carolinensis (Voigt) 1832

The green anole is one of the most conspicuous lizards in South Central Texas. It is usually bright to pale green but is capable of readily changing its color to several shades of brown. Because of this quick-change capability, it is often referred to as a "chameleon" (true chameleons are not found in North America). This slender, arboreal lizard belongs to the lizard family Iguanidae which contains some 65 genera and about 650 species, most of which are found in the Western Hemisphere. The genus *Anolis,* with over 250 species, makes up the largest genus of reptiles in the Americas. *Anolis carolinensis* is the only species of anole native to the continental United States. This anole occurs in the southeastern U.S., from southern Virginia, south to the Florida Keys, and west to the Rio Grande River in southwestern Texas.

Identification: Anoles can change color within minutes from a bright

23

green to a brown or intermediate shade. Occasionally, a light-colored or dark brown to mottle stripe is found running down the back of this reptile. A distinct feature of the male (usually reduced or absent in the female) is the expandable throat fan or dewlap. *This extendable dewlap on the male is usually of pink or reddish color.* The dorsal scales are generally small and keeled, with the exception of the slightly enlarged middorsal rows. The head is long with a pointed snout. The body is lacking a distinct dorsal crest, the tail is round, and the postanals are enlarged in the males. Characteristic of all anoles are the well-developed, adhesive toe lamellae or pads (with claws extending beyond them) that are used for climbing. Expanded toe lamellae and the presence of display fans distinguish *Anolis* from all other iguanids in Texas. Because of its distinctive green color, body shape, and habits, the green anole is not likely to be confused with other lizards in South Central Texas.

Size: Generally, adults range from 5 to 8 inches (12.7 to 20.3 cm) in length. Females are slightly smaller than males.

Behavior: Males of this diurnal, arboreal lizard can exhibit a wide variety of social displays, including head and body movements accompanied by the spreading of the dewlap. The dewlap is used during the visual signaling and is used in combination with head-bobbing, body push-ups, posturing and tail movement, and occasionally during opening of the mouth. The green anole will use its repertoire of displays during courtship, aggressive encounters, and while establishing territory. Displaying its dewlap involves the movement of a complex lever-like mechanism formed by the hyoid bone, which expands the fan down for display. Male anole lizards are strongly territorial; if an intruder continues to approach another's territory, an aggressive encounter usually ensues. Color changes in this arboreal anole usually involve several factors, i.e., just because it is on a green leaf does not mean that it will be green. The actual color changes are a result of movement of pigment in the cells of the lizard's skin. Additionally, a complex combination of factors involving light intensity or quality, temperature fluctuation, social behavior, and humidity or other environmental factors is the driving force behind the fluctuations of its color. No other lizard in South Central Texas is capable of undergoing such a drastic color change.

In this area hibernation presumably starts in mid-December through February, although they have been observed on a year-round basis, depending on the winter warming trends. In wooded areas they sometimes hibernate in areas beneath the bark of stumps or under rotten logs and occasionally are found curled up in ground depressions (Michael, 1972).

Food: Prey animals in the diet include flies, moths, mayflies, lacewings, damselflies, cockroaches, beetles, crickets, caterpillars, mealworms and other insect larvae, soft-bodied insects in general, and many types of spiders. Although they are arboreal in nature, I have occasionally observed them foraging on the ground between shrubbery or vegetated areas.

Habitat: The author, as a science enrichment teacher, asked thousands of local students if they had ever spotted a green anole in their yard. The response was overwhelming. Three out of four students raised their hand. If you provide shade and shelter, and water the lawn and plants in your yard, you have the makings of the anole's chosen environment. This lizard is commonly distributed throughout the region in areas such as yards, heavily vegetated areas adjacent to streams and ponds, forests and woodland edges, old building locations, and shrubbery along roadsides. Their use of toe pads with claws allows climbing on rubbish piles, fences, low trees, shrubs, vines, and low vegetation with some associated shade.

Breeding: This is an egg-laying species. Mating occurs in March through spring into the summer. The male's gonads are influenced by photoperiod, i.e., lengthening days of spring (Fox and Dessauer, 1958; Licht, 1973), and the female's ovarian development also is triggered during this courthship-displaying season. Females are reported to store viable sperm for a minimum of eight months after copulation with the male. *Anolis carolinensis* is reported to be capable of producing one egg every two weeks for the entire breeding season (Hamlett, 1952), and the author has observed variability in oviposition frequency over one season (confirmed by Andrews, 1985). All *Anolis* females apparently develop only a single ovarian follicle at a time, ovulation alternating between the two ovaries. One soft-shelled, white egg is deposited on soft, moist, leaf litter and humus or in a rotten log. Depending on temperature, the egg hatches in about 6 to 7 weeks, typically from July through November. Hatchlings are about $2^{1}/_{4}$ to $2^{5}/_{8}$ inches (5.8 to 6.7 cm) long. Although there is some disagreement among herpetologists, male anoles probably mature in 6 to 8 months, according to Fox and Dessauer (1958), while females mature in 12 to 15 months.

General information: At his study area along the San Antonio River, the author has made over 100 nocturnal observations of these lizards. They usually sleep on leaves or branches in shrubs or vines about 18 to 36 inches (45.6 to 91.2 cm) above ground level, although occasionally they are spotted 2.7 to 3.6 yards (3 to 4 m) up in the black willow trees along the riverbank. At night, without exception, all lizards were green when a flashlight first illuminated them. Apparently the green anole sleeps with its legs tucked against its body, which is positioned along the length of a leaf or stem.

Longevity records, according to Burrage (1964), indicate that they may live in excess of five years in captivity. Bowler (1977) lists the longevity record at over seven years; longevity in nature probably averages considerably less than five years.

Cats are known predators of these lizards. According to Pianka (pers. comm.), the rough green snake (*Opheodrys aestivus*) is also a documented predator of *Anolis*.

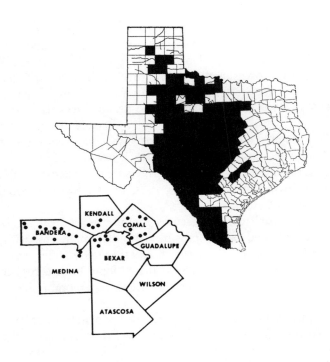

TEXAS EARLESS LIZARD

Cophosaurus texanus texanus Troschel 1850

These fast-moving, diurnal lizards have no visible external ear membranes. The Texas earless lizard is a member of the "sand lizard" group of iguanids, and there are two subspecies in Texas. Only one subspecies, *Cophosaurus texanus texanus,* occurs in South Central Texas.

Identification: This moderate-sized lizard (females tend to be smaller than males) has several distinguishing characteristics. *The conspicuous, broad, black crossbars under the flattened tail (in both sexes) offer the observer an especially fast identification.* Often when the lizard is perched on a rock, it exposes the ventral tail crossbars when curving its tail upward and "wagging" it from side to side. Dorsal bands of spots are also found on the tail, becoming larger and fused toward the tip of the tail. Lizards that have regenerated tails lack these crossbars on the regrown part. A prominent feature, more distinct in the male, is the two diag-

27

onally curved, bold, black marks on each flank just in front of the back legs, curving down from the lateral surface to about one-third of the ventral area on either side. Dorsally the light area between these dark, curved marks is occasionally yellow to light orange, and ventrally is bordered by a blue area on either side in males. Gravid females have orange spots on the sides and throat and usually lack black, bold markings.

The ground color of these lizards is variable, tending to resemble that of the substratum, i.e., soil or rock color, within different geographic areas. Color ranges from grayish to reddish-brown dorsally, marked with irregularly spaced, dark spots and many small, light flecks over most of the surface.

Other identification features include *no external ear membranes showing,* a flattened body, long hind limbs, small, smooth, granular dorsal scales, ventral scales larger than other body scales, small, dorsal head scales, small throat scales with two loose folds across the throat area, enlarged postanals, and usually 12 to 17 femoral pores present on each back leg.

Similar lizards in South Central Texas include the plateau earless lizard (*Holbrookia l. lacerata*), found in the Edwards Plateau region, and the southern earless lizard (*Holbrookia l. subcaudalis*). They superficially resemble the Texas earless lizard, except that *Holbrookia* have rounded, dark spots rather than crossbars under the tail.

Size: Adult Texas earless lizards range from about 2³/₄ to 7¹/₈ inches (7.0 to 18.4 cm) in total length, with an average snout-vent length of 2³/₄ to 3¹/₄ inches (7.0 to 8.3 cm).

Behavior: Many of these lizards have been found resting, sleeping, or hibernating, partially covered in loose shale and gravel areas of the limestone Hill Country of the Edwards Plateau. Occasionally, some specimens are found up to 6 inches (15.3 cm) under this substrate accumulation. When this lizard is disturbed from its original site or pursued by a predator, it may run directly to a rock or tree shelter, then make several trips between the two sites. According to field observations, these lizards apparently do not inhabit permanent burrows and avoid temperature extremes by moving from one area to another, as do other lizards that have wide temperature tolerances. With the possible exception of the Texas horned lizard, the Texas earless lizard seems to have a higher optimum temperature range than many other lizards in the South Central Texas area. The earless lizard appears to be mostly active between 9:00 A.M. and noon, with almost one-third of the lizards seen between 11:00 A.M. and noon.

Male earless lizards occasionally engage in aggressive body movements, showing a challenge display accompanied by push-ups when confronted by another male that has entered its territory. Personal observations indicate that this lizard may run for a distance and stop, then occasionally elevate and curl its tail upward and forward, move the tail from side to side, and reveal the distinct, black bars on the underside of its tail. This behavior often occurs when the lizard is perched on a large rock or an elevated lookout point. Tail-breakage may not play as important a role when escaping a predator as in the slower-moving species of lizards. This lizard depends on its speed as a method of escaping its predators, and has been clocked at a little under 5 feet (1.52 m) per second.

Food: The diet includes grasshoppers, crickets, small beetles and their larvae, winged termites, and spiders.

Habitat: This ground lizard lives along dry limestone ledges along streams, limestone outcrops, large boulders in river beds, and crevices along the canyon floor floodplain. Within these habitats it tends to prefer flat areas with little ground cover rather than dense grass cover. The earless lizard has been recorded in upland and riparian open areas in association with catclaw-cedar, mesquite, live oak woodlands, and along gravelly sandstone or limestone creeks. Several were once found buried on a narrow shale ledge on a bluff along a riverbank, approximately 3 to 9 feet (0.91 to 2.74 m) above the normal water level, considerably below flood level. The loose shale next to these lizards was relatively moist (but not wet) and they were simply raked out of the substrate. This bluff faced a southern exposure, with the area receiving about 7 or 8 hours of sunlight each day during the winter.

Breeding: The reproductive season, i.e., egg development in females and gonad development in males, is from the end of March to August. Courtship behavior has been observed in Central Texas during May, June, and July. This oviparous lizard has been reported to lay 3 to 8 eggs, with 5 eggs the mean clutch size. Females are reported to deposit several broods per season, e.g., 3 to 5 clutches totaling 18 to 25 eggs. These eggs are usually found singly, 1 or 2 inches (25 to 51 mm) underground. Eggs from a single clutch are usually spread throughout the area (possibly a dispersal strategy related to potential predation). Eggs measure $9/16$ inch (15 mm) by $5/16$ inch (8 mm). The incubation period is estimated at about 50 days, with hatching documented on June 29 and July 7. Hatchlings are about 2 inches (51 mm) long.

General information: Documented predators of this lizard include the Texas night snake (*Hypsiglena torquata jani*) and the western coachwhip (*Masticophis flagellum testaceus*). *Cophosaurus* also makes great food for a variety of captive snakes. The type locality for this species is New Braunfels, Texas.

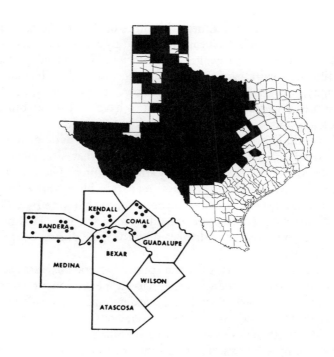

EASTERN COLLARED LIZARD

Crotaphytus collaris collaris (Say) 1823

This large, terrestrial lizard is recognized by its large head, black neck markings, and fast bipedal movement, i.e., occasionally lifting its front legs off the ground while running. This member of the family Iguanidae occasionally shows up in pet stores as it survives relatively well in captivity (provided the proper requirements of light, food, and temperature). The genus *Crotaphytus* ranges from eastern Missouri, west to California, and south to the northeastern part of the Lower California Peninsula. *Crotaphytus c. collaris* occurs in the Hill Country counties of Bandera, Kendall, Comal, and the northern sections of Bexar and Medina counties that extend onto the Edwards Plateau.

Identification: This brightly colored lizard has two conspicuous, dark collar markings that can be seen from a distance. The variable dorsal ground color is usually dull green mixed with brown and gray, with variable,

31

small, light spots scattered near the edges of the visible narrow bands or bars across the back. Color is influenced by a number of factors, i.e., sex, age, breeding, and geographic location. The dorsal markings on adult lizards usually fade as the lizards become older. The adult male lizard may display bright shades of yellow, orange, green, or blue during the breeding season. Females tend to have dull colors but develop scarlet spots and streaks along the sides when gravid.

Juvenile lizards have several well-defined, dark brown, transverse, dorsal crossbars made up of dark spots that alternate with yellowish bands for the first few weeks after hatching. The ventral side of this lizard is white. Males possess a relatively blunt throat-fan that is pouch-like in appearance and bordered on either side by tightly inflated skin that is compressed laterally along the fan. This blunt and relatively incomplete fan is displayed by the male during courtship and territorial behavior.

External characteristics of the species include *an exceptionally large head,* large, oval ear membranes, narrow neck, long tail and hind limbs, small, smooth, dorsal body scales, and tail scales noticeably larger than dorsal body scales. They average 16 to 21 femoral pores on either side; most males have enlarged postanals; and many individuals have two complete rows of interorbital scales on top of the head. No other lizard in South Central Texas has a large head, thin neck, and double black collar. The crevice spiny lizard (*Sceloporus poinsettii poinsettii*) also has a black collar, but it is single and broad and margined by light-colored scales. Moreover, members of the genus *Sceloporus* have very large, spiny dorsal scales.

Size: The total length for adults, including the tail, is about 8 to 14 inches (20.4 to 35.6 cm).

Behavior: This diurnal lizard is often found in pairs, even during the nonbreeding season. Territorial behavior of a resident male lizard may include throat-puffing, exposing a colorful yellow-orange coloration, body elevation and flattening with bobbing or jerking movements, and occasionally charging at the intruder. These behaviors are intended to drive off newcomers and usually result in their speedy retreat. Gravid females are also highly territorial. Perching behavior can be observed along the upper parts of large boulders and outcrops along open areas. Such positions apparently serve as prime lookouts for foraging, territory supervision, and thermoregulation. Hibernation sites include shallow burrows under large, flat rocks, and small, body-sized, open-ended burrows underneath large, limestone shelf rock.

These small, body-sized burrows apparently are much smaller than the larger mammalian burrows, although the latter may also be used. Hibernation usually lasts from November to late March, but fluctuates with prevailing weather conditions. Records indicate that most collared lizards are active at temperatures of 74 to 93 F. (23.5 to 34 C.). If alarmed, these lizards rapidly take cover in established safe spots and crevices, under rocks, or in holes. When captured, they will not hesitate to deliver a painful bite. If detached, the tail does not regenerate in the same manner as in most other lizards.

Food: Stomach samples and captive observations indicate that they are primarily insectivorous opportunists, eating mostly grasshoppers, beetles, cicadas, butterflies, wasps, crickets, and wolf and jumping spiders, as well as other available soft-bodied arthropods. Occasionally small vertebrates, i.e., baby cotton rats, earless lizards, small skinks and other lizards, and small snakes, are eaten. Captive collared lizards thus should not be placed in the same cage or bag with small lizards or small snakes.

Habitat: C. collaris inhabits dry, open, rocky limestone ledges and loose rocks, and along washes and outcrops of the Edwards Plateau. The habitats for this terrestrial lizard are thus limited by the Balcones Fault Zone, which separates the sedimentary formations of the rolling hills of the southeasternmost portion of the Great Plains from the coastal plains to the south.

Breeding: The breeding season is reported from April through June. Mating usually takes place in April and May, with 1 to 13 eggs per season being reported to be deposited in loose soil, with the exception of 24 (probably in error) reported by Strecker in 1910, although the average clutch size is about 5 to 9 eggs. A study by Trauth (1978) indicates that *Crotaphytus collaris* produces two clutches of eggs per season, one in April or May and the other in May or June. The usually white, cylindrical eggs measure about $3/4$ inch (19 mm) by $7/16$ inch (12 mm) and may require 51 to 86 days to hatch, depending on the temperature during incubation. Hatchlings are about $3^1/16$ to $3^5/16$ inches (7.9 to 9.2 cm) in length and usually first appear in August and September. These hatchlings may mature in one to two breeding seasons (Fitch, 1956). Hatchling growth is rapid, about half of its adult size being reached in only a few months. Female lizards apparently may become mature in the first year (Ballinger, 1985) and are reported to lay at least five eggs for the first clutch.

General information: Potential major predators include roadrunners, snakes, e.g., the Central Texas whipsnake, and hawks.

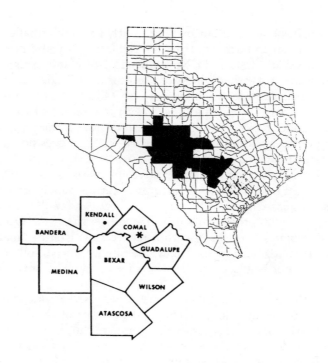

PLATEAU EARLESS LIZARD

Holbrookia lacerata lacerata Cope 1880

The plateau earless lizard is a member of the genus *Holbrookia*, considered part of the "sand lizard" group in the family Iguanidae. The species *H. lacerata* is made up of two subspecies; these differ externally only in their markings and femoral pores. *Holbrookia l. lacerata* is restricted to the Edwards Plateau and adjoining areas in Texas. Only a limited number of sightings of this Hill Country inhabitant have been documented for the South Central Texas area; these specimens are from around Helotes in Bexar County, the Guadalupe River area in Kendall County, and in Comal County.

Identification: The dorsal ground color for the plateau earless lizard is light grayish-brown. *Characteristic markings include two series of dark, often light-margined, roughly crescent-shaped markings running along the length of the body and the upper surface of the tail.* Often the dark, crescent-

34

shaped, dorsal spots are fused and band-like. *About six or seven small, round, black spots are usually present on the undersurface of the tail in both sexes.* These tail spots tend to fade posteriorly near the tip of the tail. Usually there are up to 6 small, lateral, abdominal marks near the ventral and dorsal line for this species. The ventral surface is normally white. Dorsal scales are small, flat, granular, and smooth, with the exception of some weakly keeled scales on juveniles. The ventral scales are large and the femoral pores number about 10 to 14 on each side. The body is about the same length as the cylindrical tail (in cross-section), and the legs are moderately long for its size. Like other members of the genus *Holbrookia*, the *tympanum is covered by scales.*

Similar species of earless lizards do not usually possess a series of six or seven black dots on the undersurface of the tail, usually fading toward the tip. The subspecies *Holbrookia l. subcaudalis* differs from *Holbrookia l. lacerata* by a higher average femoral pore count (averages three more than *H. l. lacerata*) and a higher percentage of specimens which have unfused pairs of blotches on each side of the vertebral line. The Texas earless lizard (*Cophosaurus texanus*) is distinguished from the plateau earless lizard (*Holbrookia l. lacerata*) by its conspicuous, broad, black bars under the flattened tail.

Size: Adult lizards range from about 4¹/₄ to 5³/₈ inches (10.8 to 13.7 cm) in length.

Behavior: During territorial challenge behavior, this lizard will raise and lower the forepart of its body a specific number of times. Push-ups begin from a crouched position and the lizard usually remains rigidly motionless, other than the up-and-down movement (Clarke, 1965). This diurnal lizard will rapidly seek shelter if approached by a predator. Courtship behavior has been documented for this species, with the male occasionally approaching the female with head lowered and back slightly arched, followed by rapid head-nodding sequence. If the female does not reject him, the male may nudge the cloacal area of the female, touch her with his tongue on the body or head, and then attempt to secure a grip on her by biting the area on the back just above her shoulder. If the female cooperates passively, the male may attempt to mount her for copulation.

The preferred activity temperature range, reported by Clarke (1965), for this species is between 99 and 104 F. (37.5 and 40 C.), e.g., cloacal temperatures between 9:00 A.M. and 5:00 P.M. during June through August when the lizards are most active. Axtell (1956) reported that they are usually most active between 10:00 A.M. and noon,

35

depending on ground or surface temperature.

Food: The diet of this earless lizard includes small grasshoppers, crickets, small beetles, other soft-bodied insects, and spiders.

Habitat: It is normally found in association with caliche soils of the Edwards Plateau, upland range and pastures near cultivated areas, and flatter areas. It avoids heavy vegetation and is most likely found in open, grassy areas in association with light, sporadic ground cover. Lizards have been found around mesquite and scattered ashe juniper (*Juniperus ashei*) in shallow ravines along dry, open areas or in low, scattered vegetation.

Breeding: Information on reproduction by this egg-laying species is limited. The female may lay two clutches of eggs per season (Axtell, 1956). Eggs may be deposited during May through August. Female lizards will vary in the number of eggs they will lay. They may lay from 4 to 12 eggs, depending on age and size, during each ovipositing period. The average length of the eggs is about $9/16$ inch (15 mm). Incubation takes about 4 to 5 weeks, with the eggs hatching in July through October. Hatchlings are about $1\frac{1}{2}$ inches (38 mm) long.

General information: Historical records indicate a widely dispersed distribution for this lizard along the southern Edwards Plateau. Although most Edwards Plateau populations of this form have now disappeared, Pianka (pers. comm.) says he saw the last ones on his property about 1971. The type locality for this subspecies is in the vicinity of Helotes, the type specimens having been collected by Gabriel Marnock in 1879.

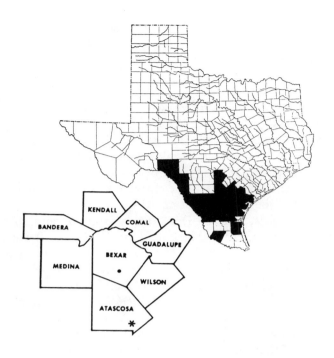

SOUTHERN EARLESS LIZARD

Holbrookia lacerata subcaudalis Axtell 1956

This small earless lizard is adapted for life in the dry mesquite-prickly pear associations of South Texas and the southern Atascosa County area in South Central Texas. It makes excellent use of its long back legs by running quickly across the surface if approached by an intruder. If an intruder grabs the tail of the lizard, it will usually fracture, making identification more difficult since the seven underside tail spots are one of the key characteristics for southern earless lizard identification.

Identification: The dorsal ground color is usually grayish-brown, marked with four rows of distinct blotches that extend along the length of the upper body and continue down the base of the tail. Furthermore, *lateral to each of these two dorsal series of markings is an adjacent row of small, dark streaks or bars,* somewhat less pronounced posteriorly as they extend along the lateral fold and disappear. This *earless* lizard's head is marked with a dark frontal spot and is bordered by two pairs of dark, transverse band markings on the supraocular region. The neck section has two irregular, elongated dark markings on each side prior to the first set of *dorsal body blotches, and the limbs are marked with light-bordered blotches.* The ventral side of this lizard is white, with the *underside of the tail marked with seven or fewer, small, round, black spots.* External characteristics include a blunt head, 15 to 17 enlarged supraoculars, small, granular dorsal scales, two gular or throat folds, extending up either side of the body and terminating slightly anterior to the forearm, 15 to 17 femoral pores on each back leg, and ventral scales approximately three times the size of the smooth, flat dorsal scales.

Similar lizards, like the Texas earless lizard (*Cophosaurus t. texanus*), have broad, squarish, black markings under the flattened tail. The plateau earless lizard (*Holbrookia l. lacerata*) differs in having 14 or fewer femoral pores, and dark blotches usually fused together in pairs producing a single row effect on either side of the lizard's back. One other lizard, the keeled earless lizard (*Holbrookia p. propinqua*), also differs in having keeled scales and no subcaudal tail spots.

Size: The maximum size for this lizard is 6 inches (15.2 cm), although 4³/₄ to 5³/₄ inches (12.1 to 14.6 cm) in length appears to be the average adult length.

Behavior: The documented behavioral actions of the species indicate that submissive males have differing attitudes and movements of the body parts, as documented by Clarke (1965). Submissive males moved away from other lizards 66 percent of the time, held the body flat 67 percent of the time, raised the tail all of the time, and moved the tail during the tail-raising procedure 84 percent of the time. Agonistic males were not noted during Clarke's study.

Food: The diet is similar to that of the plateau earless lizard, consisting of a variety of grasshoppers, crickets, small beetles, soft-bodied insects, and arachnids.

Habitat: The southern earless lizard is usually found in association with certain soil and vegetation types in South Texas. Dark clay and clay loam soils are usually the home of this dryland lizard. The habitat, which includes a mesquite-prickly pear association, generally extends along a line from Maverick County to Karnes County. *Holbrookia l. subcaudalis* seems to skirt the sandy soil habitat of the keeled earless lizard (*Holbrookia p. propinqua*), preferring the flat clay or clay loam fields in South Texas.

Breeding: Information on the reproduction of this subspecies is limited. Females may lay two clutches of eggs, one clutch in May or June and another in July or August. Approximately 4 to 12 eggs are laid, depending on the age and size of the female. The eggs average about ⁹/16 inch (15 mm) in length and hatch in about 4 to 5 weeks, usually in July through October. Hatchlings are about 1¹/2 inches (38 mm) long, approximately the same length as the plateau earless lizard hatchlings.

General information: Little is known concerning its range into Atascosa County, other than a Louisiana Tech University specimen (LTU 4570) collected near the southeastern border in Atascosa County by W. K. Davis on June 30, 1956. Axtell's (1956) distribution map shows them just south and east of Atascosa County, in Live Oak and Karnes counties. Edward Taylor collected one in southern Bexar County, indicating they were once in the county many years ago (Axtell, pers. comm.).

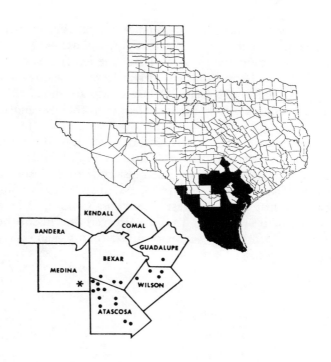

KEELED EARLESS LIZARD

Holbrookia propinqua propinqua Baird and Girard 1852

This relatively small, long-tailed lizard ranges from South Texas to northeastern Mexico. The species consists of two subspecies, one in Texas (*H. p. propinqua*) and the other in Mexico. The keeled earless lizard, although moderately common in the sandy areas of northern Atascosa, northern Wilson, and southern Bexar counties, is most abundant along the sand dunes of Padre Island. There it grows slightly larger in size than its mainland counterpart in South Texas.

Identification: The dorsal ground coloration for this species is often a uniform tannish with a grayish or olive tint. Dorsal markings are often with dark, posteriorly light-rimmed spots. Some specimens have an unmarked middorsal area that runs down the body, usually bordered by dorsal blotches that continue down onto the tail, where they become smaller and mesh together toward the tip of the tail. Numerous small, whitish

40

or occasionally light yellow spots are found along the dorsal blotches. *Two distinct, diagonal, dark lateral bars, present behind the forelegs on the male lizard,* allow for easy field identification and sex determination. Female lizards have indistinct or variable markings. One gravid female in southern Bexar County was lemon-yellow on the dorsal sides from head to tail and striking dark orange to red on the lower dorsal sides and the underside of the tail. Males develop a pink coloration on the underside of the tail and a variously colored throat spot. External characteristics for identification include *no visible ear membranes,* two folds across the throat, small, granular, slightly keeled dorsal scales, ventrals larger than dorsal scales, enlarged postanals on males, and *10 to 21 femoral pores (14 to 16) on each side.*

The only other similar lizard found within the range of the keeled earless lizard is the southern earless lizard *(Holbrookia lacerata subcaudalis). H. l. subcaudalis* differs in having the underside of the tail marked with seven or fewer small, rounded, black spots, and in having relatively flat, smooth, granular dorsal scales.

Size: The total length of this lizard as an adult in South Central Texas is about 4 to 4³/₄ inches (10.2 to 12.1 cm), with the barrier island females (e.g., Padre Island) averaging a little larger, up to about 5¹/₂ inches (14.0 cm) in length.

Behavior: The keeled earless lizard forages opportunistically for insects along open, sandy areas around shrubs. It may pursue flying insects, and has been observed climbing into low shrubs to capture prey. According to Judd (1976), *H. propinqua* "seemingly eats ants rather 'passively' as a consequence of usual behavior involved in movements."

Courtship observations by Cooper (1985) indicated that males court nonresident females more intensely than local resident females. This study also indicated that males can distinguish resident from nonresident females. Most *H. propinqua* lizards restrict their movements to a limited area, possibly facilitating escape from predators. The mode of escape for males is usually a prolonged run, with the female making a relatively short dash to vegetation. Field experiments by Cooper (1986) suggest that a territorial male will respond to the orange and yellow components of female secondary sexual coloration. By identifying a resident female in her relatively small territory, the yellow may curtail agonistic approaches by the dominant territorial male. Orange is displayed to males by females actively rejecting courtship (Cooper, 1986), although this coloration may stimulate courtship activity in other lizards like *Crotaphytus* (Axtell, pers. comm.).

41

Food: An analysis by Judd (1976) of 129 adult and 45 juvenile stomachs indicates that grasshoppers, beetles, beetle larvae, and spiders constitute the most common food items, totaling 76 percent of the food items taken. Other food items in the diet include crickets, ants, flies, and leafhoppers; a mantid was found in one stomach. Several hatchling stomachs contained only ants, and no prey larger than $1/8$ inch (3 mm) was found in any hatchling.

Habitat: This lizard is found on the Carrizo Sands Formation of southern Bexar County, and in Atascosa and Wilson counties. It is observed in open loose sand deposits interspersed with grass clumps, low vegetation, and shrubs throughout their range. On the Gulf barrier islands off the coast of Texas and Mexico, it is found near the edge of sand dunes. When kept in captivity, the keeled earless lizard may quickly shimmy down and disappear into the loose sand of the terrarium.

Breeding: The reproductive season in South Texas may extend from late March through August, with hatchlings appearing during June. The clutch size variation in one population of keeled earless lizards from mainland South Texas ranged from 3 to 7 eggs, varying from year to year depending on environmental conditions (Judd and Ross, 1978). These lizard eggs measure $7/16$ inch (12 mm) by $1/4$ inch (7 mm). Moreover, there is indirect evidence that females may lay an average of four clutches of eggs each year (Judd, 1976). Females appear to become sexually mature at about 10 months and males may mature in only 9 months. Hatchlings are about $1 1/2$ inches (38 mm) long and are marked with a distinct pattern of paired blotches.

General information: The type locality for *Holbrookia propinqua* is listed as "Indianola to San Antonio." John H. Clark, an assistant to Lt. Col. J. D. Graham of the U.S.–Mexican Boundary Commission, probably collected the original syntypes as he crossed the Carrizo Sands Formation southeast of San Antonio in April or May of 1851 (Axtell, 1981).

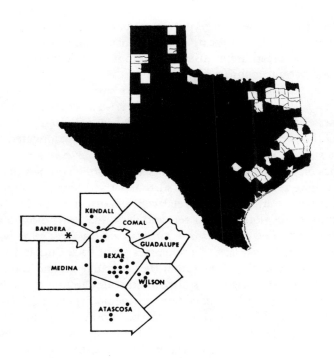

TEXAS HORNED LIZARD

Phrynosoma cornutum (Harlan) 1825

The genus *Phrynosoma* consists of about 14 known species, ranging from southern Canada to Guatemala. *Phrynosoma cornutum,* the only species represented in South Central Texas, ranges from Kansas to northern Mexico. The Texas horned lizard surely lives up to its name, as it possesses an imposing array of daggar-like spines on the head, sharp ridges over the eyes, and pointed fringe scales on either side of the abdomen. The term "horned toad" (used in the past and occasionally today) is a misnomer. Compounding the problem, the generic name *Phrynosoma* literally translated means toad-body, as at a distance this wide, flat-bodied creature may look more like a toad than a lizard. This species, formerly sold in the pet trade, is now protected by state law from exploitation. Anyone who collects one and takes it home is violating state law. But the drastic reduction of its distribution and

43

population density seen in recent years in South Central Texas was not primarily caused by collecting pressure but by some unresolved factor, perhaps widespread use of insecticides (indirect evidence) or major agricultural habitat modification.

Identification: The variable ground color ranges from a light yellowish-brown to shades of tan, reddish-brown, or gray. The color pattern usually consists of a distinct, whitish, middorsal stripe extending from the nape of the neck to the tail, and two dark brownish, elongated spots on either side of the neck. Additionally, four or more round, dark blotches on either side of the middorsal line (often rimmed to the back with a light yellowish or white) surround the enlarged dorsal spines. Other pattern features include relatively less distinct, rounded or fused markings on the dorsal perimeter, and dark lines extending downward from each eye. *This unique lizard is armed with two large horns on the back of the head.* These two large center spines point backward and are much larger than any of the other spines that continue around the fringe of the head. There are two rows of pointed scales around the sides of its broad, flat body. Other characteristics include short legs, visible ear membranes, a distinct series of enlarged spines on each side of the throat, small, keeled dorsal scales with enlarged keeled scales on the middorsal area, and keeled ventral scales. No similar species are found in the area.

Size: The average size for an adult is about 4 to $5^{1}/_{8}$ inches (10.2 to 13.1 cm) in total length, although a record $7^{1}/_{8}$-inch (18.1 cm) lizard was found in Big Bend National Park in June 1969.

Behavior: This diurnal lizard is usually active in the midmorning until noon, and again in the late afternoon, thus usually avoiding the hottest part of the day. Temperatures during these time periods usually range within this lizard's optimum body temperature of 100 F. (38 C.). If temperatures are excessive, the horned lizard may bury itself by pushing its snout into the sand with side to side movements while its body is also moving in a similar fashion. The substrate fills in around the lizard as it moves, exposing only its eyes and nostrils. For protection, the horned lizard usually depends on its camouflaging coloration of dull, sandy gray that blends in with its habitat. Another common defensive behavior when threatened involves projecting the horns on its head vertically, inflating the lungs, or expanding the ribs out to look relatively flattened. Moreover, horned lizards possess the unusual ability to increase blood pressure in the head and force out blood from the anterior corner of each eye. The behavioral aspects of forcing blood from its eyes was reported by Collins (1982). This rare behavior usually happens during stress-related situations, and possibly as part of the lizard's skin-shedding process. Lambert (1985) reported that of 97 *Phrynosoma cornutum* encounters only 7 ejected blood when they were first encountered. Cowles (1977) reported blood ejection by the fringe-toed sand lizard (*Uma notata*), another similarly adapted iguanid species, indicating that this behavior may occur in other forms.

This lizard hibernates in late October or early November and reappears in late March or early April in South Central Texas. When stalking its prey it makes quick forward thrusts, bending the head forward as the rapidly moving tongue captures the unsuspecting insect.

Food: The principal food of this lizard is the "harvester" ant *Pogonomyrmex* and other large ants, but they are not exclusively ant-eaters. Other insects included in its diet are small ground beetles, weevils, grasshoppers, crickets, and true bugs. Due to the relatively specialized diet of this lizard, most Texas horned lizards are unlikely to experience severe competition for food items from other lizard species in the area.

Habitat: This terrestrial lizard can be found in South Central Texas along almost any type of open, flat, semiarid terrain with limited grassy areas including floodplains, cropland, rangeland, pastures, and upland rocky ledges. The soil association is variable including loamy, rocky, or sandy loose soil types. Usually mesquite, cactus, and bunchgrass are found in the general area of the home range. Vegetation, burrows, or rocky areas provide limited refuge for this lizard, and it occasionally burrows in loose soil under bunchgrass or shrubs.

Breeding: The breeding season for this species is from April until mid-July. Females may produce only one clutch of eggs per year (Ballinger, 1974). Clutch size ranges from 14 to 37 eggs per clutch, with the average at 29 eggs (Ballinger, 1974) or 23 eggs (Fitch, 1985). These creamy-white, tough, leathery eggs usually measure 7/16 inch (12 mm) by 5/8 inch (16 mm). The female lays her eggs in a burrow dug to a depth of 5 to 9 inches (12.7 to 22.9 cm), usually at a 45- to 75-degree angle from the horizontal entrance. Within the nest, eggs are usually deposited with the soil filled in around them. Newly hatched lizards usually emerge 39 to 47 days after deposition (in August and September, depending on the temperature), and measure just under 1 inch (25 mm) in length. Females apparently mature in their second season after birth. Observed mating of this lizard indicates that the male may grasp, within its mouth, one of the female's large head spines during copulation.

General information: Snakes and birds, e.g., hawks, roadrunners and ravens, have been documented eating Texas horned lizards. Natural predation, however, is not responsible for this lizard's major decline in the last 20 years. Commercial exploitation as pets from the late 1950s, before their protection, probably had limited impact on the overall population dynamics of the species. Moreover, they seldom make good captives, as their requirements include live ants and exposure to full sunlight. Instead, several reasons have been put forth as alternative explanations for the spectacular disappearance of these lizards in recent years. Included among these are epidemic disease, pesticide contamination, extensive modifications of habitat for agriculture, and destruction of their prey base by fire ants. Nevertheless, none of these competing explanations are as yet supported by anything but circumstantial evidence.

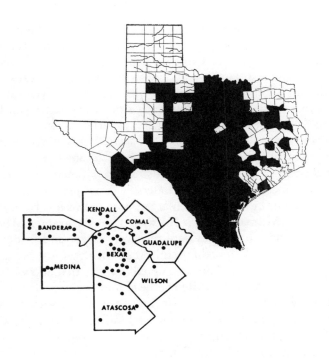

TEXAS SPINY LIZARD

Sceloporus olivaceus Smith 1934

This abundant, mostly arboreal, diurnal lizard is a member of the genus *Sceloporus,* a large North and Central American group of approximately 100 species. This genus has 13 subspecies in Texas and 4 recognized species in South Central Texas, making it the best represented lizard genus in the area. One of the two large species of the spiny lizard is found from south-central Oklahoma through South Central Texas to northeastern Mexico.

Identification: The dorsal ground color of this relatively large, rusty-brown spiny lizard is predominantly brown with a variable mixture of gray, black, and white. Male lizards of this species generally have a moderately distinct, light-colored dorsolateral stripe and feature a narrow blue patch on each side of the pale ventral surface. These bright blue patches extend from the forelimbs to the hind limbs and increase in width and

47

brightness with age. Females can be distinguished by the presence of numerous, distinct, dark gray, transverse, wavy bands across the dorsal area. The dorsolateral light stripes in females are usually not very well-defined; the undersurface is paler, with small spots on variable dark gray. Although some females lack the blue color on the ventral surface, many have a faint blue patch on each side. Both sexes can become darker to favor heat absorption and paler to reflect heat.

This lizard has *large, heavily keeled, pointed dorsal scales,* numbering from approximately 28 to 33 from the head to above the vent at the base of the tail. Caudal scales are also strongly keeled and resemble stiff, sharp spines, except for the underside scales just behind the vent. Males have a pair of noticeably enlarged postanal scales. Femoral pores range from approximately 11 to 16, with 13 as the average. The 5 or 6 supraocular scales on the head are relatively large, and in one series. Specialized toes are designed for climbing, with each toe being tipped by a sharp claw. The toe claws are curved in a way that allows the lizard to hang on the tree bark while the head and body are facing down, slightly away from the main vertical tree trunk. Another identifying characteristic is that there is no gular fold across the throat, a feature that is found in the (much smaller) genus *Urosaurus.*

Similar species in South Central Texas include the crevice spiny lizard (*Sceloporus p. poinsettii*), distinguished by a dark collar band across the neck, the southern prairie lizard (*Sceloporus undulatus consobrinus*), and the northern fence lizard (*Sceloporus undulatus hyacinthinus*), distinguished by their smaller adult size and smaller dorsal scales (35 or more from the back of the head to the base of the tail). The northern fence lizard usually has a dark line running along the rear surface of its thigh. The rosebelly lizard (*Sceloporus variabilis marmoratus*) is distinguished by its smaller adult size, smaller dorsal scales, and two large, pink ventral patches.

Size: Average size for females, three years and older, is 9 to 11 inches (22.9 to 27.9 cm) in total length. Males of comparable age are smaller than females, with males averaging only $7^{1}/_{2}$ to 9 inches (19.1 to 22.9 cm) long. Newborn hatchlings average about $2^{1}/_{2}$ to $2^{5}/_{8}$ inches (63 to 67 mm) at birth.

Behavior: Texas spiny lizards are primarily an arboreal species. Their markings blend well with the tree bark, and they warily avoid detection by moving to keep the trunk of the tree between them and potential predators. This "squirrel-like" behavior allows them to be overlooked even in areas where they are numerous. The spiny lizard occasionally descends to the ground to forage for food, to move to another tree, or to lay its eggs. According to Blair (1960), who did extensive work on this species, nearly all basking, some feeding, and most mating is accomplished in the trees or on elevated objects such as walls, posts, or fences. Movement of individual adult spiny lizards, according to Blair's study, ranged over an area of about 0.07 to 0.16 acres.

The ground-foraging behavior usually involves the lizard spotting an insect on the ground, leaving its position on the tree, rushing to the ground to capture the insect, and then returning shortly to the same tree. If threatened, they will rapidly climb the tree to a safe overlook, or, if on open ground, rush into a large hole (such as an armadillo burrow). Blair (1960) took the cloacal temperature of several adult lizards that were basking and recorded an average temperature of approximately 97 F. (36 C.). The lowest temperature recorded was 89 F. (31.5 C.) at 8:45 A.M., and the highest was close to 103 F. (39.5 C.) recorded at 4:00 P.M. These lizards are occasionally active during winter on warm, sunny days. The author has observed them on several occasions basking at air temperatures of around 80 F. (29 C.) in December, January, and February. Most individuals hibernate at some time during the winter.

Food: The Texas spiny lizard, like nearly all other North American lizards, feeds primarily on insects and some other varieties of arthropods. Although grasshoppers constitute a substantial portion of the diet, other insects such as blister beetles, June beetles, Lepidoptera larvae, "pill bugs" (order Isopoda), and such arachnids as spiders and mites are taken. According to stomach content analyses, flying insects are seldom eaten. Additionally, Wayne McAlister reported one unusual instance of a Texas spiny lizard eating a small 8³/₄-inch (22.3 cm) rough green snake (*Opheodrys aestivus*).

Habitat: This lizard is abundant in mesquite savannas and occurs in association with hackberry, live oak, elm, pecan, juniper, cottonwood, willow, persimmon, and other trees. Additionally important habitats include large prickly pear clumps, old buildings, abandoned homes, stumps, logs, bridges, fences, and rock walls which feature shelter in the crevices. Hibernation is known to occur under the cover of heavy leaf litter and soil.

Breeding: Mating usually begins in mid-March in South Central Texas and continues through early summer. The author has observed a nest site, along an elevated slope of loose soil, being excavated at about a 45-degree angle to a depth of 3 or 4 inches (7.2 or 10.2 cm). After the eggs were laid, they were covered with soil. This observation of egg-laying was made in late April, with other authors reporting additional observations in April, June, July, and August. Depending upon age, 1 or 2 (possibly up to 4, according to Blair, 1960) clutches of eggs are laid per year, extending the egg-laying season through the summer. Females may become sexually mature during their first reproductive season of the second year, laying an average of about 11 eggs in the first reproductive season, and an average of 24 eggs per brood in their third season thereafter. The number of eggs per clutch depends on the size and age of the female. Males may mature at the onset of their first breeding season in March of the second year. It was estimated by Blair (1960) that only about 2 to 5 percent of the eggs laid would hatch and the young reach sexual maturity.

General information: Another common name for the Texas spiny lizard is the rusty lizard. Predators include the Texas patch-nosed snake, western coachwhip, Schott's whipsnake, young Texas rat snake, broad-banded copperhead, armadillo, skunk, house cat, opossum, roadrunner, and other large birds. Blair (1960) estimated that patch-nosed snake predation pressure is probably very important in relationship to its reproductive ecology. Small red mites are ectoparasites, often found on the groin or axillary area of this and many other lizards. Estimated maximal life span is about five years.